Food
78

欢迎每个人

Everyone Is Welcome

Gunter Pauli

[比] 冈特·鲍利 著

[哥伦] 凯瑟琳娜·巴赫 绘

李欢欢 牛玲娟 译

上海远东出版社

丛书编委会

主　任：田成川

副主任：何家振　闫世东　林　玉

委　员：李原原　翟致信　靳增江　史国鹏　梁雅丽
　　　　任泽林　陈　卫　薛　梅　王　岢　郑循如
　　　　彭　勇　王梦雨

特别感谢以下热心人士对童书工作的支持：

匡志强　宋小华　解　东　厉　云　李　婧　庞英元
李　阳　刘　丹　冯家宝　熊彩虹　罗淑怡　旷　婉
杨　荣　刘学振　何圣霖　廖清州　谭燕宁　王　征
李　杰　韦小宏　欧　亮　陈强林　陈　果　寿颖慧
罗　佳　傅　俊　白永喆　戴　虹

目录

Contents

早晨，小猪正在农场里散步，他停下来和他的朋友们聊天。

"早上好，昨晚睡得好吗？"小猪问。

"嗯，挺好的，谢谢！"一只母鸡说，"你把这里收拾得很温馨舒适！整个房间暖融融的！和你住在同一个屋檐下，真好！"

$\mathcal{A}$ pig is taking a morning stroll around the farm and stops to talk to his friends.

"$\mathcal{G}$ood morning. Did you sleep well last night?" asks the pig.

"$\mathcal{O}$h yes, thank you!" says one of the hens. "You've made it so cosy for us here! You warmed up the whole room. How good it is to live under the same roof."

住在同一个屋檐下，真好！

How good it is to live under one roof!

你们可能会把流感传染给我

you could give me the flu

"很高兴为大家服务。不过，我很多朋友认为我疯了才和一群鸡住一起。他们说你们可能会把流感传染给我。"

"嗯，我们不用挤在小笼子里，呼吸自己的粪便浮尘，而是睡在你上面，享受你身体散发的热量。同时你保护我们的安全，防止狐狸进来。多放松啊。"

"My pleasure. However, most of my friends think I'm crazy to sleep in the same place with chickens. They say you could give me the flu."

"Well, we are not packed tightly into small cages, breathing the dust of our own manure. Instead, we sleep on a perch above you to enjoy your body heat while you keep us safe by keeping the foxes at bay. It is so relaxing."

"你们吃掉了所有烦人的虫子和苍蝇，帮助我保持清洁和健康。你不介意帮我捉身上的寄生虫，对吧？那感觉就像温柔的按摩。"

"和你住一起，真好！你瞧，你吃的东西，我们不吃。你吃剩的残渣养活了土里的生物，这些生物让我们更强壮。"母鸡说。

"And you help me stay clean and healthy by eating all the worms and flies that bother me. You don't mind picking a few parasites off my skin, do you? It feels like a gentle massage."

"It is so great to have you in our community. You see, what you eat, we don't eat, and the droppings you leave behind feeds creatures in the soil that make us so much stronger," says the hen.

让我们更强壮

Make us so much stronger

有我们生存所需的所有东西

We have all we need to live

"的确是这样。这周围有森林、灌木丛和草地，有我们生存所需的所有东西。"

"但我们当中有些鸡并没能活很长时间。"母鸡突然伤心地说。

"That's true. Around here, with the forest, the bush, and the meadow we have all we need to live."

"But some of us do not get a chance to live long," says the hen, suddenly sad.

"谁想杀害你们？记住了，我会赶走所有的狐狸！"

"不，不是的。我不担心那些捕食者。我是担心那些人类，他们总是想要更快地得到更多的食物。"

"哦，我明白了！没错，人类压力很大，他们觉得有很多穷人会死于饥饿，除非企业能生产更多更便宜的食物。"

"Who wants to kill you? Remember, I will keep all foxes away!"

"No, no, it's not our predators that I'm worried about. I'm worried about those people who always want more and more food faster and faster."

"Oh, I see. Yes, people are stressed, as they believe that many poor people may die of hunger unless businesses produce more and cheaper food."

谁想杀害你们？

Who wants to kill you?

是因为你的儿子们不能下蛋吗?

Is that because your boys do not lay eggs?

"专家声称为了能更高效地生产更多食物，一旦我的儿子们孵化出来，全都得被杀死。"

"是因为你的儿子们不能下蛋吗？我猜想主营鸡蛋销售的人认为养公鸡没用。"

"These experts claim that to make more food more efficiently, you need to kill all my boys as soon as they hatch."

"Is that because your boys do not lay eggs? I suppose that people whose main business is selling eggs think that there is no use in keeping the males."

"那正是养鸡企业不再需要我的兄弟们，想要全部杀了他们的原因。"

"嗯，欢迎你所有的兄弟们来这里，他们在这里可以过着幸福健康的生活。"

"非常感激你的提议。但你知道吗？同样的事情也发生在我女儿们的身上。"母鸡说。

"That's exactly why the chicken industry does not want our brothers anymore and want to kill them all."

"Well, all your brothers are welcome here where they can lead a happy and a healthy life."

"We are so grateful for that. But did you know the same happens to my girls?" asks the hen.

同样的事情也发生在我女儿们的身上

The same happens to my girls

不能长出和公鸡一样多的鸡肉

Not able to produce as much meat as the boys

"我以为他们会留着母鸡下蛋呢。"

"唉，他们会留下一些品种中的所有母鸡来下蛋。但另一种用来产肉的鸡就不一样了。一旦那些鸡孵出来了，人们会把母鸡和公鸡分开，杀死所有的母鸡，因为她们不能长出和公鸡一样多的鸡肉。"

"I thought they keep the girls to lay eggs?"

"Well, for some kinds of chickens they do keep all the hens for laying eggs. But it is not the same for another kind of chicken, those reared to produce meat. Once those chicks hatch, people separate the boys from the girls. All the girls are then killed, as they are not able to produce as much meat as the boys. "

"所有这些都是以提高产量为名义吗？让母鸡们也都来这里和我们住在一起吧。"

"谢谢你如此关心并邀请别人分享你的住所。我们确实生活在富足的土地上，在这里大自然永远都为我们提供充足的食物。"

……这仅仅是开始！……

"And all this is going on in the name of productivity? Let the girl chicks come live here with us as well."

"Thank you for being so caring and inviting others to share your space. We certainly live in the land of plenty where nature's table is always laden with food."

... AND IT HAS ONLY JUST BEGUN!...

……这仅仅是开始！……

... AND IT HAS ONLY JUST BEGUN! ...

Did You Know ?

你知道吗?

For every female, egg-laying chicken that hatches, a day-old male chicken is killed. For every male chicken that will produce meat that hatches, a day-old female chicken is killed.

每孵化一只下蛋的母鸡，一只刚孵化的公鸡就会被杀害；每孵出一只产肉的公鸡，一只刚孵化的母鸡就会被杀害。

It takes only four weeks to fatten a rooster with soy, corn, and power food, including hormones and antibiotics. A chicken eating grass, seeds, and insects would need four months to reach maturity.

用大豆、玉米和含有激素以及抗生素的强力饲料养肥一只公鸡只需四个星期，一只吃草、种子和虫子的小鸡需要四个月才能长成熟。

4 周

4 个月

A "super" hen lays around 330 eggs per year and will survive for a maximum of 18 months. A "natural", domesticated hen would lay 220 eggs per year.

一只"超级"母鸡每年产出约 330 枚鸡蛋，最多只能存活 18 个月。一只自然家养的母鸡每年能产出约 220 枚鸡蛋。

All specialised egg-laying and meat-producing chickens are inbred to the point that they are genetically degenerate and cannot procreate anymore. All 50 billion chickens that are produced industrially each year are the product of artificial insemination.

专门产蛋和专门产肉的鸡是近亲繁殖，基因退化，无法再生育。每年批量孵化的 500 亿只小鸡是人工授精的结果。

Pigs are living in perfect symbiosis with the forest. They get food from the forest without destroying it, while making the forest fire resistant by keeping the undergrowth in check.

猪和森林是完美的共生关系。猪不破坏森林，从森林获得食物，同时控制灌木生长以防森林着火。

One thousand hectares of forest, fields, and bush can host 10,000 chickens, one thousand pigs and one thousand sheep without any need to provide them with commercial feed.

1000 公顷森林、田地和灌木可以养活 10 000 只鸡、1 000 头猪和 1 000 只羊，无需提供任何人工饲料。

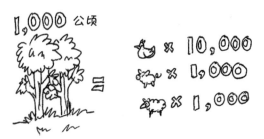

Pigs were domesticated from wild boars in China 10,000 years ago. They were originally tuber-eating forest dwellers. Pigs do not sweat and therefore cannot cool their bodies down. That is why they live under the canopy of trees and enjoy rolling in mud.

猪是由中国人在1万年前驯养野猪而来，猪最初是食用块茎的森林居民。猪不流汗，因此无法让身体降温。这也是它们住在树底下，喜欢在泥里打滚的原因。

Pigs are very smart and are ranked more intelligent than dogs.Every year two billion piglets are fattened to 200-kg hogs in only six months.

猪很聪明，比狗还机智。每年，仅在6个月内就有20亿只小猪被养到200千克。

Do you agree that as the world needs more food all female chickens should be removed from meat production and all male chickens from the fertilisation process?

因为这个世界需要更多的食物，就要在肉类生产中杀害所有母鸡，在鸡蛋生产中杀害所有公鸡，你赞同吗？

你认为生活在森林里的动物需要外界的食物吗？

Do you think that animals living in a forest will need any imported food?

Which system will produce more and healthier food and replenish topsoil: the one where only chickens are raised or the one many animals are raised together, alongside a variety of plants?

只养殖鸡的系统和同时养殖多种动植物的系统，哪种能产出更多更健康的食物并补充表层土？

生产更多的相同产品是最好的解决方法吗？或者，我们能同时生产多种食物，而不是只生产越来越多的鸡蛋吗？

Is producing more of the same always the best solution? Or, instead of producing only more and more eggs, could we have several sources of food produced at the same time?

Have you ever visited a chicken farm? (We are often kept from visiting modern battery chicken farms because of the risk of contamination, as you may become infected and get sick.) Calculate how many eggs you eat per day and then per week. Also look at the labels of any packaged and processed foods you have in your house as many of these contain eggs. Extend you research to also calculate how many eggs other members of your family consume. How many chickens in total are needed to supply your household with eggs? Now have a look at how much kitchen waste and leftover food there is available every day. Do you think that it would be enough to feed chickens of your own?

参观过养鸡场吗？（我们通常不参观现代化养鸡场，因为有污染风险，也许会感染生病。）计算一下你每天吃几个鸡蛋，每周吃几个。再看看你家里食品包装袋的产品信息，有多少食物含有鸡蛋。家里其他成员消耗多少鸡蛋？共需要多少只鸡才能满足家里的鸡蛋需求？现在看看每天有多少厨房垃圾和残羹剩饭，你认为这足以养活你家所需要的鸡吗？

学科知识
Academic Knowledge

生物学	植物蛋白相对鸡肉、猪肉和牛肉动物蛋白的转化率;有草原、灌木和森林的生态系统的再生能力;人类起源于热带,如今我们模仿最初的住处来设计保温的房子;共生关系,动植物如何相互协作来创造高效系统;鸟类控制自然虫害。
化 学	鸡的粪便富含氮(4%)、磷(2%)和钾(1%);家禽粪便的pH值随着年龄和食物变化,维持在6.5至8之间。
物 理	热空气会上升,因此,猪散发的热量上升至鸡笼的地板,给鸡提供温暖,这是生物气候学的房屋设计基础。
工程学	猪有逃跑的倾向,因此需要智能门系统把猪圈起来;猪、鸡、鸭和羊共生场地的设计。
经济学	把鸡养在笼子里以提高产量;每增加1千克体重,猪需要吃掉2.5千克食物;人口增长和食物产量息息相关,需要提高产量。
伦理学	健康食物和食物量之间的平衡;生命和产量之间的选择。
历 史	2 500年前,赫拉克利特已讨论生命的和谐,指出所有事物都受制于养分、物质和能量之间的流动;大约9 000年前,中国和印度已驯养小鸡。
地 理	土耳其位于欧洲和亚洲的交界处。第一次世界大战后,庞大的奥斯曼帝国瓦解,现在的土耳其是其中的一部分。
数 学	土地面积固定,人口增长,所以食物产量必须提高;每平方米需要80至400条蚯蚓才能保证土壤健康,这就意味着每英亩地需要超过100万条蚯蚓。
生活方式	我们的饮食越来越依赖于动物产品,似乎没有人意识到如果我们摄入更多的素食,能更快地解决世界饥饿问题;给流浪者、孤儿和难民提供庇护的文化。
社会学	多元文化社会的生活与相同阶级、相同种族和相同文化人群一起生活的对比;语言形成了对雄性、雌性和幼崽的特定称呼。
心理学	拥有足够的安全感,能睡得更香甜;放松程度影响脑海中的信息。
系统论	共生的重要性:共同协作制造更多养分、物质和能量。

情感智慧
Emotional Intelligence

小 猪

小猪对自己身边的每一个人都怀有深深的同情心，关心他人的幸福。即使存在禽流感的威胁，他还是希望和鸡住在一起。小猪感到庆幸，优雅地解释他如何给鸡提供温暖，保护他们远离捕食者。他意识到应对世界饥饿问题需要更多食物，那就需要多元化生态系统。为了产量，杀害公鸡或母鸡，他认为这不合理，并非常同情他们，愿意给所有不被接受的小鸡提供住所，甚至欢迎两倍数量的鸡（公鸡和母鸡）来这里。

母 鸡

母鸡对小猪给予她的关心很高兴。对于小猪提供的温暖和保护，母鸡感到幸福快乐并表达了自己的感激之情。她坦率地讨论起伤心的话题，说出了自己对其他鸡的担忧，他们没有机会像她一样感受生活。母鸡明确地回答了小猪的问题和疑虑，揭开了问题背后的真实原因。当小猪说欢迎其他人来这片富足的土地时，母鸡感到兴奋。

艺术
The Arts

你在鸡蛋上画过画吗？拿三个煮熟的鸡蛋，在一个鸡蛋上画棵树，在另一个鸡蛋上画只鸡，在第三个鸡蛋上画一只猪，试着顺着鸡蛋的曲线画出它们的脸。记住可以用白色或棕色作为鸡蛋的底色。如果喜欢，你可以画更多的鸡蛋！

思维拓展
Systems: Making the Connections

社会不停地寻找提高食物产量的方法，一方面希望解决世界饥饿问题，另一方面希望获取更多的利润。我们似乎完全认识到了食物生产对动物生存质量的巨大影响。稀缺资源的有效管理带来了复杂的喂养系统，为了达到更高的生产水平，喂养过程中的每一步都经过详细研究。让我们想想人类食物生产行为如何影响和我们一起生活在地球上的其他物种的生存质量，这些行为看似没有人性。因为性别原因无法达到生产的最高水平，所以雄性蛋鸡和雌性肉鸡被杀害了。这种杀害不仅限于鸡类，还有牛类，因为商人只关注牛奶的产量。动物们更愿意和谐地一起生活在食物富足的地方，这样会带来多样化：多种动物有多种养分来源，提供了强大的共生关系基础。比起单纯地关心环境，这还需要改变我们的计算方式和商业模式。通过作出改变，我们能不断地创造、支持和维护共生关系，补充土壤，加强生态系统，以便后代能收获超乎我们想象的大量优质产品。所有这一切使我们能够在地球有限的资源下可持续地生存。

动手能力
Capacity to Implement

我们首先要意识到食用肉类或蛋类是个人选择。我们作为消费者，要求高品质食品很重要。只有消费者需求，高品质食品才会被生产，政府才会制定食品生产法则。食品生产商通常只会关注降低成本，赚取利润。我们的购买力很大程度上决定了能否实现改变。如果成千上万的消费者要求高品质鸡蛋，那么就会生产出这样的鸡蛋。如果现有的企业不能生产这样的鸡蛋，那就给创业者一个机会满足市场。和家人朋友探讨一下，告诉他们你会为杀死公鸡和母鸡的现状感到伤心。看看周围有多少人支持你，坚信鸡应该被养大，改变杀害幼鸡的现状也能创造足够的食物。

故事灵感来自
This Fable Is Inspired by

格奥尔格·施魏斯富特
Georg Schweisfurth

格奥尔格·施魏斯富特在德国长大，生活在欧洲最大的腊肠加工厂之一的附近，那里每天加工牛的数量多达 5 000 头。他是一个屠夫，在慕尼黑和弗莱堡学习商科。比起他父亲的食品生产方式，他坚信应该有更好的方式。《互联思维艺术》作者弗雷德尔克·韦斯特、永续生活设计的共同创立者比尔·莫里森、《一根稻草的革命》作者福冈正信，启发他 1988 年在德国慕尼黑东南的格隆镇成立了 Hermannsdorfer Landwerkstätte。之后，他还创立了 EPOS（一切皆有可能）组织 Gut Sonnenhausen 生态酒店和 Basic AG 连锁生物超市。他和家人致力于给竞争激烈的农业生产和肉类加工制定新的质量标准。他著有《清醒的安德斯》一书。

图书在版编目（CIP）数据

冈特生态童书.第三辑修订版：全36册：汉英对照 /
（比）冈特·鲍利著；（哥伦）凯瑟琳娜·巴赫绘；
何家振等译.—上海：上海远东出版社，2022
书名原文：Gunter's Fables
ISBN 978-7-5476-1850-9

Ⅰ.①冈… Ⅱ.①冈… ②凯… ③何… Ⅲ.①生态环
境–环境保护–儿童读物—汉、英 Ⅳ.①X171.1–49

中国版本图书馆CIP数据核字（2022）第163904号
著作权合同登记号图字09-2022-0637号

策　　划 张　蓉
责任编辑 程云琦
封面设计 魏　来　李　廉

冈特生态童书
欢迎每个人
[比]冈特·鲍利　著
[哥伦]凯瑟琳娜·巴赫　绘
李欢欢　牛玲娟　译

记得要和身边的小朋友分享环保知识哦！
八喜冰淇淋祝你成为环保小使者！